YOUR KNOWLEDGE HAS VALUE

- We will publish your bachelor's and
  master's thesis, essays and papers

- Your own eBook and book -
  sold worldwide in all relevant shops

- Earn money with each sale

Upload your text at www.GRIN.com
and publish for free

Bibliographic information published by the German National Library:

The German National Library lists this publication in the National Bibliography; detailed bibliographic data are available on the Internet at http://dnb.dnb.de .

Imprint:

Copyright © 2015 GRIN Verlag, Open Publishing GmbH
Print and binding: Books on Demand GmbH, Norderstedt Germany
ISBN: 978-3-668-13799-8

This book at GRIN:

http://www.grin.com/en/e-book/315416/securing-personal-health-records-in-the-cloud-by-using-attribute-based

Mayur Oza, Nikita Gorasia, Nishant Doshi

Securing personal health records in the cloud by using attribute based encryption. A review

GRIN Publishing

SECURING PERSONAL HEALTH RECORD ON CLOUD USING ATTRIBUTE BASED ENCRYPTION: A REVIEW

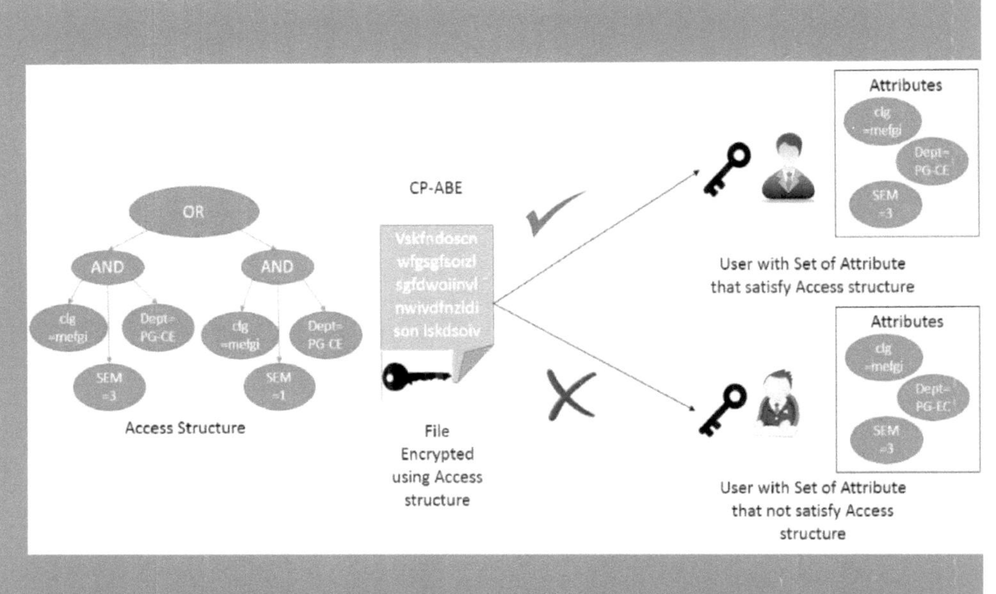

Mayur Oza, Nishant Doshi, Nikita Gorasia

MARWADI EDUCATION FOUNDATION, RAJKOT

SECURING PERSONAL HEALTH RECORD ON CLOUD USING ATTRIBUTE BASED ENCRYPTION: A REVIEW

Submitted By

MAYUR NAVNITLAL OZA

Under the Guidance of
NIKITA GORASIA and NISHANT DOSHI
Department of Computer Engineering
Marwadi Education Foundation Group of Institutions

A Dissertation Phase-1* Submitted to

Computer Engineering Department
Marwadi Education Foundation Group of Institutions
Affiliated to Gujarat Technological University

For the Partial Fulfillment of Degree of Master of Engineering
In Computer Engineering

December – 2015

Faculty of P.G. Studies & Res. in Engineering & Technology

At & PO : Gauridad, Rajkot-Morbi Road

Rajkot 360 003. Gujarat. India.

*Report is modified and updated based on Dissertation Phase-1 Presentation and comments.

Index

List of Figure

Figure no.	Name	Page No.

List of Tables

Acknowledgements

I am grateful to numerous local and global "peers" who have contributed towards shaping this Dissertation Phase 1 report.

First and foremost I would like to express my sincere thanks to *Dr. Nishant Doshi* who created my interest in *cryptography*. He was always there to guide, motivate and support me whenever I was stuck. He constantly reminded me to achieve my goal. His observations and comments helped me to establish the overall direction of the research and to move forward with investigation in depth. Irrespective of his busy schedule, he always gave time to listen to my doubts patiently and gave valuable suggestions like a parent. His doors were always open to discuss my doubts anytime.

I owe my deep sense of gratitude to Dissertation Phase 1 report examiners *Dr. Sarang Pande, Dr. Nitul Dutta* for their valuable suggestions and critical comments during presentation of credit as well as progress seminars and also sharing their knowledge which influenced me more to carry out this research work. I am thankful to *Prof. Yogesh Ramani*, for helping us to solve typos and grammatical error throughout the book.

I thank to all my student colleagues for providing fun filled and very informative environment to learn and grow. It is their love and encouragement, which helped me a lot during my research work. I had many memorable moments with them inside and out-side my work. I am grateful to all my *colleagues, M.Tech students, Teaching Assistants* and many others, for being with me in my difficult times and for all the emotional support, care, and fun they provided and who have been kind enough to advise and help in their respective roles. I owe a dept of gratitude to all my friends for their guidance and support. *Manish Shingala and Kishan Makadia* for their thoughtful discussion related to research work that helped me to complete this work in timely fashion. *Harshit Champaneri, Dhara Patoliya, Ruchita Kaneria and Jinita Tamboli* for constant motivation to carry out this research work.

I wish to thank staff of *Department of Computer Engineering, MEFGI* for providing me resources throughout my stay in the college.

Last and most important, I thank my family members. Without their constant support, motivation and love, I would not have been reached so far. I dedicate this research work to my family.

Mayur Oza

Abstract

In emerging world of cloud computing gives wide range of functionalities. Personal Health Record (PHR) enables patients to store, share, and access personal health data in centralized way that it can be accessible from anywhere and anytime. However combining of PHR with cloud gives new horizons for medical fields to be digitalized but it comes with major concern as security. There are many researchers work in securing PHR which stored in cloud using naïve approaches but it's not enough to secure it. So there is need for new technology as Attribute Based Encryption that secure PHR with providing many functionalities such as revocation of user, delegation of other user access, accountability, searching over encrypted files, multi-authority and many more. So here we survey on this field of securing PHR on cloud using ABE.

Keywords: Personal Health Records, Attribute Based Encryption, PHR on cloud, Ciphertext policy Attribute Based Encryption.

1. Introduction[1]

1.1 Traditional Health Records and Electronic Health Records (EHR):

In older days the health records are stored in medical journals/notes and manage by hospitals, but the management of hardcopy is tedious to write, share and search for some records. However the emergence of the digitalization of this medical records are converted into digital copies which is known as Electronic Health Records(EHRs) and its similarly manage by hospitals so instead of searching of records are easy, patient has no control on it. But increase cost to stores more data needs more data centers and webservers. The problem with medical data is sharing of it. Sharing of traditional data is not easy because that managed by hospitals not by patient. Data storages not able to elastically stretchable according to come with new data, however data is stored on cloud gives that service well with Database-as-a-service.

Fig. 1.1.1 Traditional Health records in older days

Fig. 1.1.2 Electronic form of traditional health records (EHR)

---

[1] The content of this book is based on the suggestions received in Dissertation Phase I presentation and report towards master degree of Mr. Mayur Oza.

1.2   Personal Health Records:

PHR are created, managed and controled by Patient itself. PHR is electronic data which includes patient's electronic health records. PHR allows user to store, retrieve and share medical data with friends, family or doctors. PHR is stored in centralized way so it can be easily accessible from anywhere and anytime. But health data is sensitive, so improper disclosure of PHR can put patient in danger. Definition of PHR by M. Li [2] as " PHR is so called because it is patients who maintain and manage these health records, that include medical records of professional diagnoses, voluntary health care programs, and other applications and services related to self-health management". Another definition as in Markle Foundation report [18] as "The PHR is an Internet-based set of tools that allows people to access and coordinate their lifelong health information and make appropriate parts of it available to those who need it."

Although, access to health data in the professional medical domain is tightly controlled by existing legislations, such as the U.S. Health Insurance Portability and Accountability Act (HIPAA) [19]. In 1996, the Health Insurance Portability and Accountability Act (HIPAA) [19] given legal privacy and security protection for PHR, but it's unable to address all issues involved because HIPAA only applies to ccvered entities.

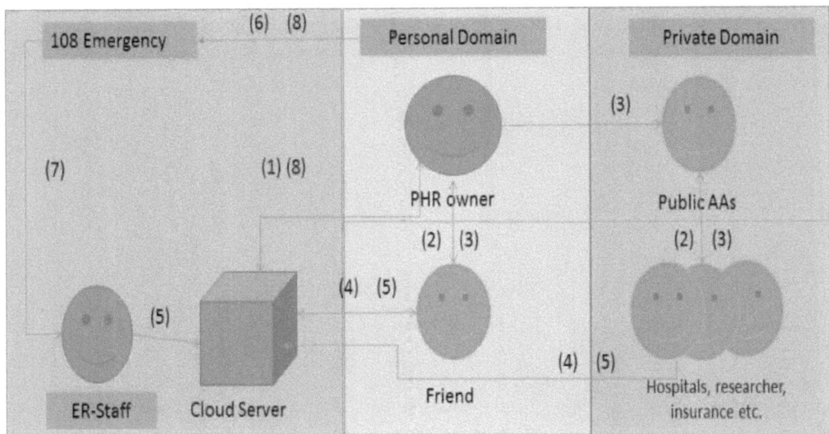

Fig. 1.2.1 Person Health Records

2   Challenges for Previous Approach and Requirement for CP-ABE

For securing PHRs on cloud may come with different flavor of encryption techniques
1. Symmetric encryption,
2. Public key encryption,
3. Identity based encryption
4. Attribute based encryption.

2.1   Symmetric Key Cryptography

Symmetric Key cryptography takes same key for encryption and decryption. Encryption algorithm takes message M and shared secret key S as input and gives output cipher text C. For decryption of ciphertext C receiver needs shared secret key S and ciphertext C so he gets original message M.

Example: Let's take an example for symmetric key encryption with two user as A and B. As describe above the both users share some secret key that use for all encryption and decryption process. As figure 2.1.1 shows sender A has Plain text M and encrypt M using shared secret key from encryption algorithm. Now this encrypted data which is ciphertext C is sent over untrusted channel. On receiving this ciphertext B gets apply shared secret key to decryption algorithm that goes successful decryption of C and gets original massage M back. But the Encryption and decryption algorithm and shared secret key must be shared between A and B before they start sharing data.

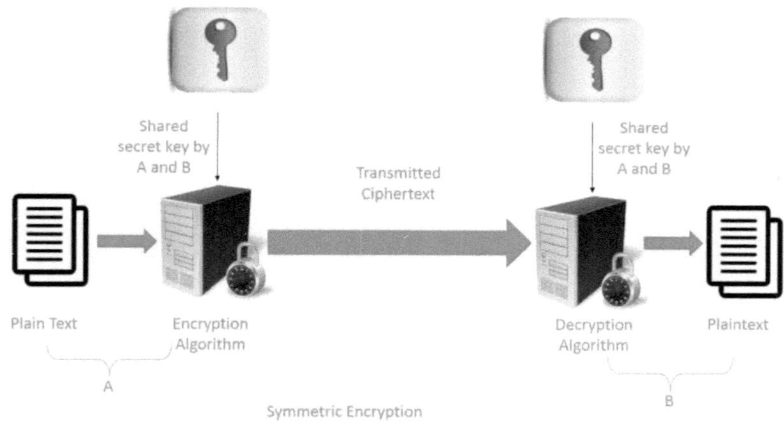

Fig. 2.1.1 Symmetric Key Cryptographic.

Limitation:

SKE is useful for smaller network as network size increase it's hard to manage key management. Main difficulty in SKE is key because the key is secret between user or group of user and if we increase size of group then chances of misbehaving user is increase and if we create new key then burden for user to store that all keys. So need for some more efficient scheme for larger network.

2.2 Public Key Cryptography

Public key cryptography takes two different keys for encryption and decryption. These key is known as public key and private key pair. So we can use anyone as encryption and other as decryption key but these gives different results and depend on our requirements.

Example: Here we consider same scenario that use in symmetric encryption where two user A and B. As describe above in public key encryption technique there is pair of keys as known public key and private key pair. Here A has two keys as PU_A PR_A and B has $PU_B PR_B$ for encryption and decryption. But here there is two scenario.

(1) PKC with private key: In these case where A encrypt plaintext using private key PR_A and send to receiver and receiver gets original massage back from using public key of A PU_A . These case also known as Digital Signature because only A can generate this encryption that can be decrypted using public key of A. Advantage of these scheme is user can encrypt data without knowing receiver.

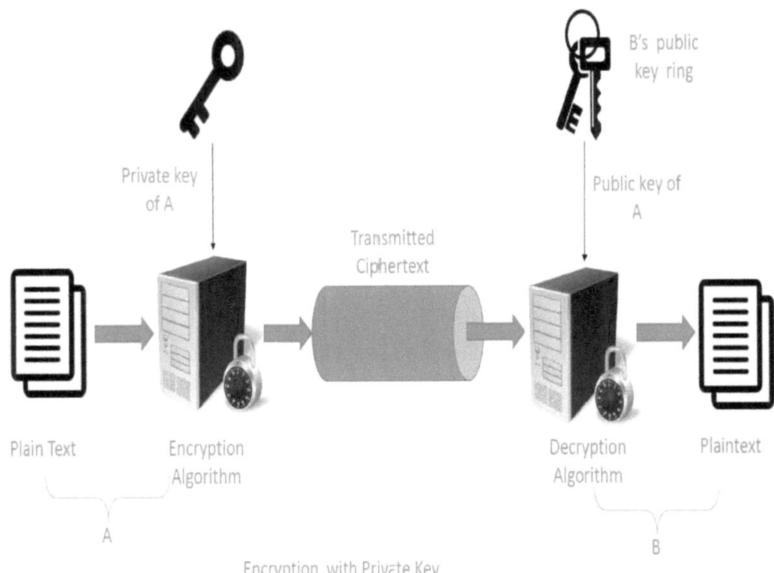

Fig. 2.2.1 Public key cryptography encryption with private key.

(2) PKC with public key : in these case message is encrypted using public key of receiver means A encrypt M using PU_B for getting ciphertext C and receiver gets message M

back in original form C by using private key of B PR_B. In these case the known as public key encryption where user has specially send massage to B so data confidentiality is archive by these case.

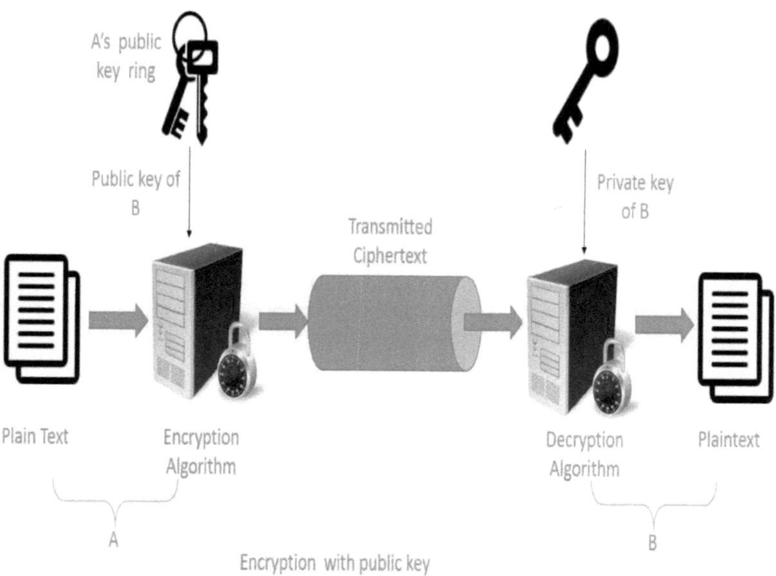

Fig. 2.2.2 Public key cryptography encryption with public key.

Limitations:

For generation of public key, private key pair need some Central Authority (CA). So there is possibility of bottleneck problem when there is more users. So we need more efficient scheme for larger network like internet.

2.3   Identity based Cryptography

To overcome limitations of Public Key Cryptography the new scheme Identity based Cryptography comes with similar approach to PKC. In IBC CA is not generating pair of public key and private key because the users identity is work as public key and user need to generate private key itself from Private Key Generator (PKG). So burden of generating two is reduce to one key.

Example: Let us take example of two user A and B and both know public key of each other. One Private Key generation (PKG) is shown in figure 2.2.3 which created complete secret on the bases of hid identity. Let A want to send data to B they encrypt data using public key of B and on other hand when B gets encrypted data they contact to PKG and generates key for his identity for decryption key. After getting decryption key they decrypt data using it.

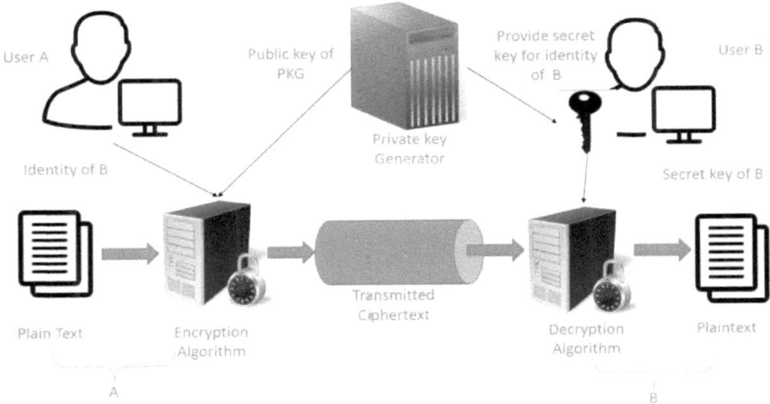

Fig. 2.3.1 Identity based Cryptographic.

Limitations: Instead of many advantages there are lots of limitations.

a) Key escrow problem: Private Key is generated by PKG so PKG can anytime use this key and encrypt message and forge sign as valid sender. If PKG is compromise then whole system is breakdown.

b) PKG is generated key for given identity so there is a chance of forge identity use by user to get private key for encryption.

c) Key Revocation: whenever user sense that Private key is compromise then they needed new identity or key for further encryption.

2.4   Attribute Based Encryption

Attribute Based Encryption is an extended work to IBE where identity of user contains descriptive attributes rather than string as in IBE. In ABE user has identity as w attributes data is encrypted using w' attributes. So when user want to decrypt this data then the attributes w' and w need some threshold level d (predefined) of similarities before that he can't decrypt that data.

In ABE there is two variants based on placing attributes and attribute policy (access structure).

    (a) CP-ABE
    (b) KP-ABE

Example: Let us consider new example for ABE scheme. Here Professor of computer branch of Marwadi college want sent some files only to students of computer branch whose study in semester 1st or 3rd in Marwadi College. So Professsor use CP-ABE scheme where required access structure is added in ciphertext so for decrypting file receiver must have all attributes that stored in ciphertext.

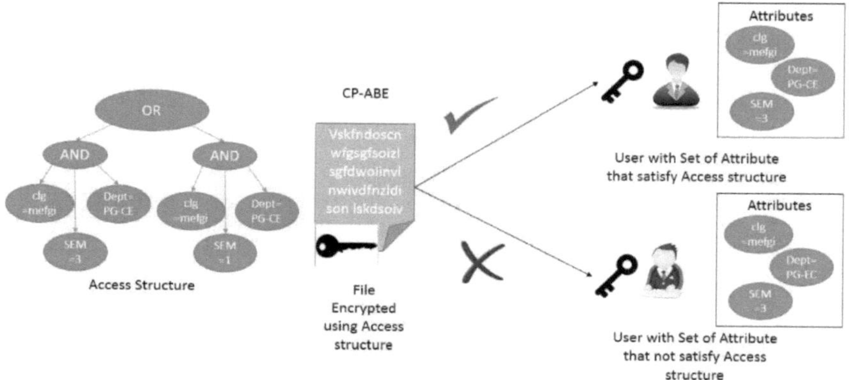

Fig. 2.4.1 Attribute based encryption ciphertext policy

3 Literature Survey

3.1 Working on PHR:

In Benaloh et.al[9] gives challenges for preserving privacy in electronic health record systems. They provides structure that gives encryption on electronic health records by encryption and also provide functionalities for sharing and searching with using keys. They formalized the patient centric encryption (PCE) for medical records and its gives advantages over electronic health records.

In Benaloh et.al [9] author proposed scheme PCE with hierarchical encryption that can be share to caregivers, doctors, and family member. They also proposed a scheme for searching over encrypted data by first encrypting the searched query and later send to server for data.

In Benaloh et.al [9] author use public key encryption and symmetric key encryption both techniques with advantages and disadvantages.

In this section, we demonstrate the code for different type of operation. For this, we use windows 7 Ultimate 64-bit operating system which have Intel core i5 2nd generation processor and 4 GB RAM and eclipse tool. And same operations are perform on Samsung Android Galaxy Grand I9082 Smartphone which have Jellybean 4.1.2 operating system, 1 GB RAM and Dual-core 1.2 GHz Cortex-A9 processor and use AIDE tool for get timing of operation.

Table 3.1.1 Properties of the schemes presented in Benaloh et.al [9]

Properties	Public key PCE	Symmetric key PCE	Flexible PCE
Upload without key distribution	Yes	No	No
Flexible Hierarchies	No	No	Yes
High Efficiency	No	Yes	No
Easy to Add categories	Yes	Yes	No

In Atallah et.al. [10] author gives problem on Benaloh et.al [9] that requires more key management and space complexity and proposed new scheme to overcome all problem on scheme Benaloh et.al [9]. Security of Atallah et.al.[10] is relies on pseudorandom function because they use pseudorandom function for generating keys. They use identity based encryption in hierarchical manners. Author proposed scheme every node has label as identity and randomly selected value as secret key. Key for lower level on hierarchy of nodes are generated by simple hashing of parents public and private information.

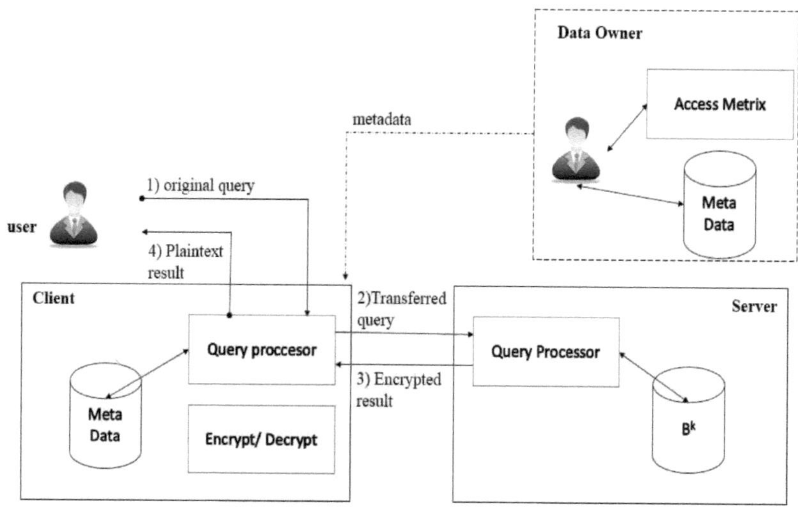

Fig. 3.1.1 DAS scenario in Damiani et.al [11]

In Damiani et.al [11] author proposed enhanced scheme for key management and encryption on selective portion. These scheme also support efficient searching over encryption. On that work author gives new model for data storage on cloud names Database-as-a-service.

In Damiani et.al [11] scheme contains portion of searching on encrypted data by using Identity based encryption. The process of searching is done as follows:
- Each query is map to encrypted query
- Sent to server and result related to query is encrypted set of tuples.
- Data of encrypted entities to client and decrypted and remove spurious tuples.
- Finally result is sent to user.

In Damiani et.al [11] users are divided into hierarchies which can be graphically presented as

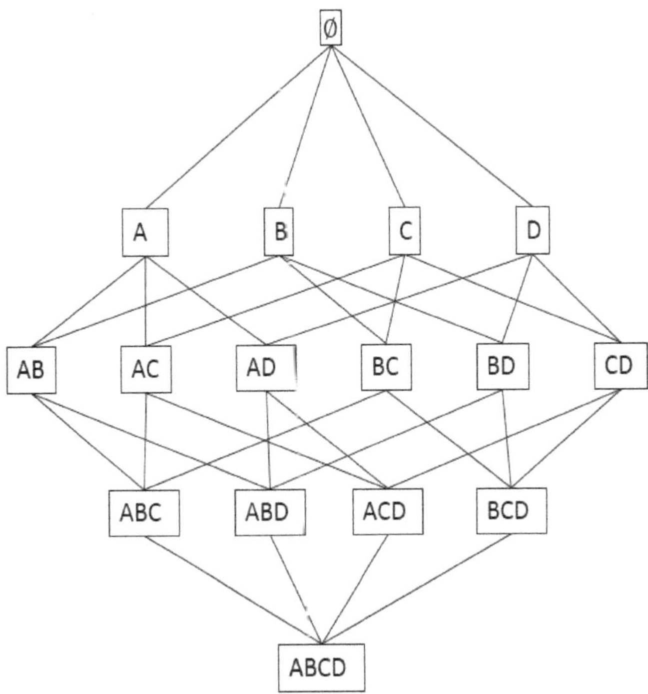

Fig. 3.1.2 An Example of User Hierarchy in scheme Damiani et.al [11].

In Wang et.al. [12] Proposed new mechanism to solve problem of owner-write-user-read applications. They encrypt every data block by different key for better security that's they achieve cryptography based access control. They use hashing function for limited computational overhead. In their scheme they introduce lazy revocation for preventing revoked user from accessing updated data block.

They gives some requirements for creating such system with this scheme are as follows:

(a) Data owner gets charges for storing data so data block should not be replicated more that's cause problem for owner.

(b) Service provider may or may not provide services on over-encrypted portion of data blocks when its send to end users.

(c) Data owner reduce number of access to provider in updating data block.

(d) In communication, storage and computational overhead for data owner and end user must be reduce.

This scheme use hierarchical key where every key in hierarchy calculated from parent node and some public information. In key derivation process hierarchy that use is shown in figure 3.1.3.

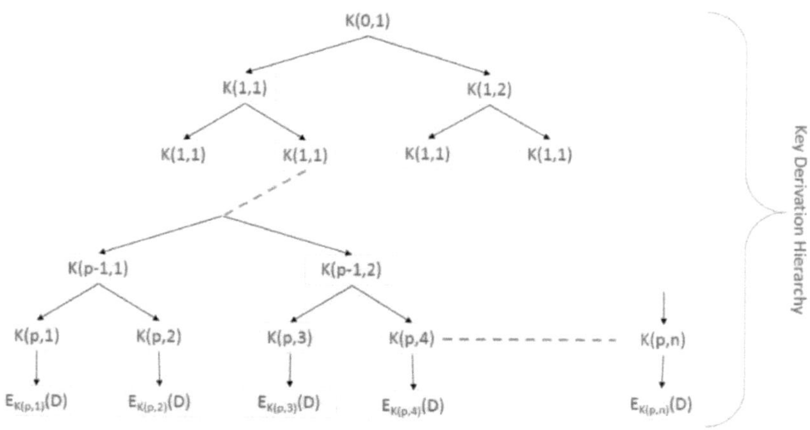

Fig. 3.1.3 Key Derivation Hierarchy in Wang et.al [12].

This scheme they shows communication overhead for giving direct access rights to end users and this comes with many problems. They give two solutions for this problem.

(a) Group data blocks with similar access pattern and give index number on outsourcing.

(b) Construct multiple hierarchies over the data block.

In analysis part of the scheme author take an example of 1Peta Byte (PB) data, 4 Kilo Byte (kb) block size, 256-bit hash and 1GHz processing power CPU for this scheme and they calculated overhead on communication part which is given in table [3.1.1].

Table 3.1.2 Overhead of the schemes presented in Wang et.al [12]

Computational overhead(in machine cycle)			
	Owner O	Server S	User U
Key-derivation	27 M	-	720 M
One-time pad generation and over-encryption	-	10 G	10G

Communication overhead			
	Owner O	Server S	User U
Data bulk index#	6 KB	-	10 KByte
Control bulk	-	10 G	10.5 KByte
Keys in hierarchy	16KB		
Updated data bulk	-	1MByte	

In Boldyreva et.al.[13] Proposed new scheme for improving key-update efficiency on the side of trusted party and efficient on end user. They reduce limitations on revocation and improve efficiency.

3.2   Working on ABE

First paper published on Attribute based encryption techniques is Sahai. et.al. [6] in fuzzy IBE scheme where identity view as set of descriptive attributes. In their scheme user with private key for identity w to decrypt cipher text which is encrypted using identityw', if and only if both identity w & w' are close enough to "set overlap" or distance metric. In their work they provide two applications for use this fuzzy IBE as follow:

(a) Fuzzy IBE system that use for biometric identity. In biometric identity where users biometric can be view as identity which describe several attribute and then encrypt to data of user using their biometric identity.

(b) On Attribute based Encryption: In ABE one party will encrypt document to all user that have certain set of attributes. Advantages of this Fuzzy IBE on ABE is that data can be store on untrusted server for everyone to share but only if they have set of attributes. Author use Shamir's method of secret sharing for distributing share of a master secret in exponents for user's private key component.

They also claims that scheme is error-tolerant, secure against collusion attack. And model is secure against selective-ID security model.

In Waters[16] proposed realization of ciphertext-Attribute based Encryption using formulas and structures. They also provide first system model allows an encryption algorithm that specify an access formula in term of any access formula. In there they express access control by linear secret sharing scheme (LSSS) matrix M.

In Bethencourt et.al.[7] Provide first implementation of CP-ABE scheme and also gives a performance analysis for given scheme.

In their scheme user's private key will associate with arbitrary number of attributes express as string. As describe in introduction, when owner encrypt message in scheme they specify some access structure with set of attributes. End user will decrypt cipher text if that user's attribute are match with cipher text's access structure.

In their mathematical level, access structure is given as monatomic Access tree, where access structure are composed of threshold gates and leaves as attributes. They use AND gate as n of n threshold gates and OR gates as 1 of n threshold gates. On comparing to CP-ABE scheme encryptor must be intelligent for knows to who can decrypt the data.   They also provide collusion resistance means multiple user collude, but they should only be able to decrypt a cipher text if at least one of them could decrypt it on their own.

As in implementation part in their work introduce new toolkit for CP-ABE construction. This implementation use pairing based cryptography (PBC) library. CP-ABE toolkit is command line tools where mainly use following commands:

They also claims that scheme is error-tolerant, secure against collusion attack and model is secure against selective-ID security model.

(a)   cpabe-setup this command generates master key and public key.

(b)   cpabe-keygen this command given with master key and generates private key for set of attributes.

(c)   cpabe-enc this command with public key and file name under access structure to specify policy.

(d)   cpabe-dec this command with private key of end user and ciphertext that what they want to access and get result on base of matching with access structure.

In Melissa et.al.[8] Introduce new approach of multi-authority that means multiple attribute authorities monitors different set of attributes and issue corresponding decryption key to user on his set of attributes before decrypting massage.

They proposed scheme that remove central authorities that requires for monitoring all Attribute Authorities which use in previous approaches for multi-authorities. They identifies two problem for multiauthority:

(a) protecting user's privacy

(b) removing a trusted authority

They presented scheme that solve this two problem for multi-authority ABE with user privacy and without trusted authority. These scheme is secure against collusion attack without concern of how many Attribute Authorities are compromised.

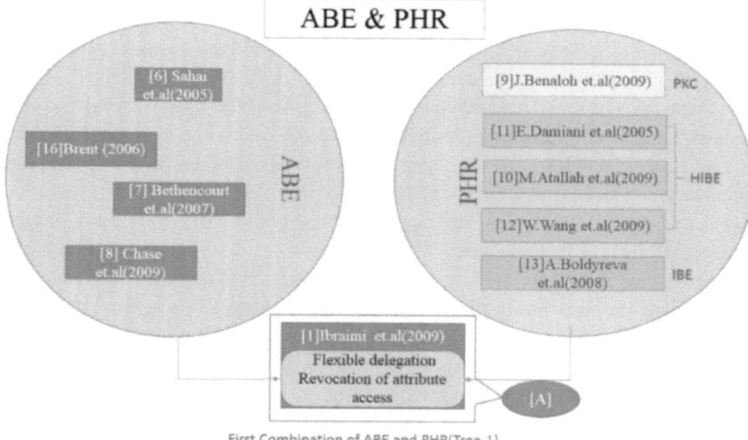

Fig. 3.2.1 Research work until combination of ABE with PHRs.

3.3   Combine Approach

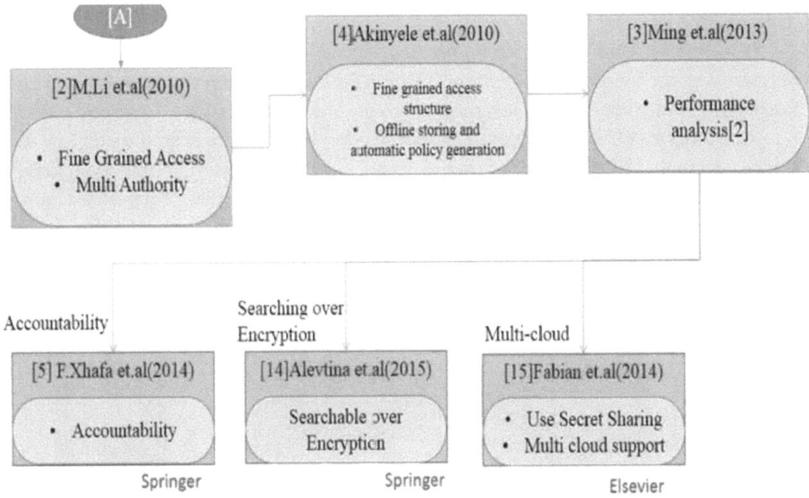

Fig. 3.3.1 Research work after combination of ABE with PHRs.

As shown in figure 3.2.1 the first time two approaches of Attribute-based encryption and PHR in L.Ibraimi et.al [1] and then work is continuously extended by adding some new functionality or give security in previous approaches.

In L.Ibraimi et.al [1] author proposed enhancement on ABE scheme using threshold decryption and flexible attribute delegation and instantaneously attribute revocation. They called these scheme as Cipertext attribute based Threshold Decryption (CP-ABTD). Advantages of these scheme are as follows:

(a) Delegator delegates his authority to delegate
(b) Delegator can decide that delegate can further delegates or not
(c) Instantaneous attribute revocation
(d) Overcome problem with uncontrolled delegation.

In L.Ibraimi et.al [1] these scheme involves three parties for encryption and accessing encrypted data (1) Trusted Authority (2) End user and (3) Mediator. These scheme divide secret key into two parts, where first part for mediator and another part for end user.to manage end users mediator has Attribute Revocation List (ARL) and Attribute Delegation List (ADL) which helps to fast and efficient delegation and revocation of unauthorized or revoked user. Trusted Authority use master key to generate secret key which is divided into two shares. Before this work the delegation of access rights to others is uncontrolled in term for further delegation is not controlled. In these scheme they give proper scenario for delegatee to restrict for more delegation from himself. So in these sense they improve delegation portion of ABE scheme. And for revocation previous schemes some time use time frame to revoked every user but in these scheme revocation gets by simply improve by adding new actor a mediator that's manages revoked user by ARL.

In L.Ibraimi et.al [1] author prove that's the scheme is secure under generic group model, where proof for generic group model is that t is based on discrete logarithm and Diffie-Hellman problem are very hard to solve as long as the order of the group is large prime number.

Author gives efficiency analysis in terms of key sizes where shared secret key size is depended on number of attributes the user have and group element in $\mathbb{G}_0$. Size of ciphertext is depend on access policy or number of attributes in access structure. They also claim that scheme is secure on collision resistance means user cannot combine their attribute to decrypt text that require attributes where any one of group has not all attributes.

In L.Ibraimi et.al [1] author use this for securing Personal Health Records that store on untrusted servers. General architecture for their scenario is given in fig. 3.3.2.

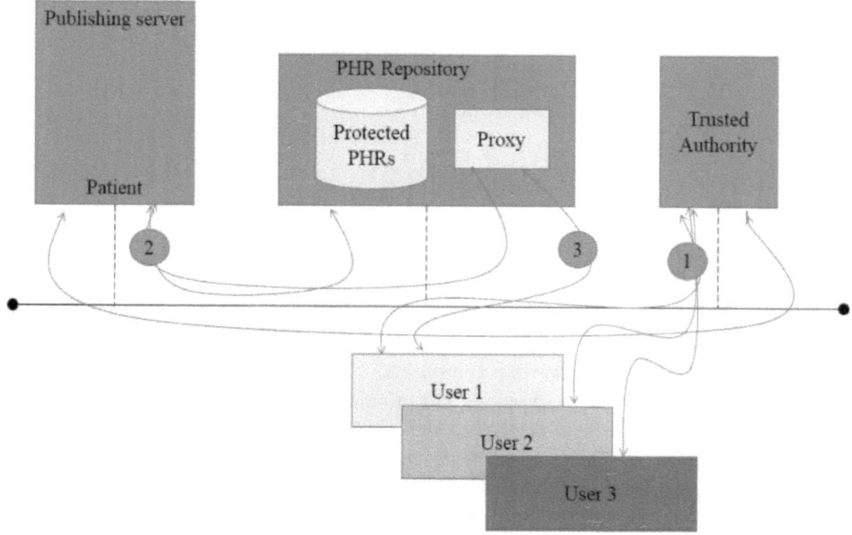

Fig. 3.3.2. Secure Management of PHR L.Ibraimi et.al [1].

As shown in figure 3.3.2 PHR has three actors and its working as follows:

(a)   Publishing server: It stored at patient side or trusted service provider. Its role is to secure and publish health records.

(b)   Data repositories: its store encrypted files.

(c)   Proxy (Security mediator): it used in data consumption state for revocation and delegation.

In M. Li et.al [2]. Mainly focused for securing PHRs which is stored at cloud using CP-ABE that gives fine grained and scalable access control. Here Authors give problem of single owner system and proposed multi-authority scheme that give list of secret key to every authority according to his attributes. Their scheme gives rights to person for encrypting file according to set of attributes. So the complexity of system is depended on number of attributes not on number of users in system.

M. Li et.al [2] Authors reduce burden of key management procedure by dividing system into different security domains (SDs) where each one has associated with subset of all users.

There each person have his friends, family members, caretaker are in personal domain and for all other person outside his circle are in public domain, which are managed by Attribute Authorities (AAs). They also add efficient and on-demand attribute revocation of user/access and break-glass access for emergency cases. Security domains are same as in fig 1.2.1. In these scheme author get solution for untrusted data storage and data encryption technique for patient centric policy by two parts. First, for lower the complexity of encryption and user management each owner of data use ABE scheme as encryption techniques. Second, divide users in two security domains as (1) personal domain (2) public domain. In these scheme owner is incharge of file inside private domain while on outside/public domains are managed by many Public AAs.

In these schemes end user who want to access data only needs to obtains credentials from the corresponding public AA and there is no need to connect to PHR owner, so here this scheme reduced the key management overhead. And for revoking user as on demanded by owner only need to update the ciphertext. Scheme is collision resistant against AA and revoked user.

In Akinyele et.al.[4] Provide design and implementation of self-protecting of electronic medical records EMR using ABE. They secure data to storing on untrusted data center. They design such a system that provide fine-grained encryption and is able to protect individual items within EMR. In their implementation part they include iPhone application for storing and managing EMRs offline that allows user for flexible and automatic policy generation. They give a problem areas to handle EMR are as follows (1) Access control: access control is main concern for large organization/hospital despite the high level of regulation surrounding use of EMR. (2) Self-protecting: data is secure at transport level for storing it but not all protected or they are bulky encrypted. (3) Complexity of access control mechanism: in large increasing world of digitization the actors and user are increasing exponential. So needs better and easy access policy for access the data need is more important. (4) EMR access is only online: access control authorities' needs to always online.

They give framework for granular role-base and content-based access controls without single, centralized server. They provide library that implement new ABE and optimization of these scheme for mobile devices. They created iPhone application that interface with GoogleHealth. In these scheme they use expressive operators in access policy which is major strength of ABE. In ABE access policies can be represented with AND, OR, NOT, threshold gates like $1 - of - n$ or $n - of - n$ and also support Boolean operators like$\leq, \geq, < or >$. Their scheme is support dual-policy ABE. Fig [] shows diagram for workflow of their scheme.

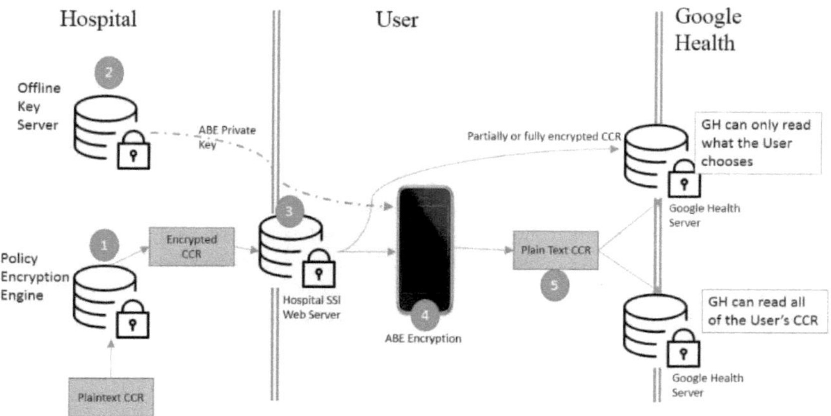

Fig. 3.3.3. Scheme of Akinyele et.al [4]

Explanation of diagram as follows:

(1)   Policy encryption engine: it parses XML-based record to appropriate access policy after that if any node match access control rules, then encrypted records using dual-policy ABE with appropriate access policy.

(2)   Once records has been encrypted, this encrypted data or stored within owner's server.

(3)   May be exported to semi-trusted cloud –base storages.

(4)   For locally stored data can be accessing by private key generation.as shown in figure.

In Xhafa et.al [5] author gives scheme that provide all functionalities of M. Li et.al [2] [3] with new ability of accountability. They use multi-authority CP-ABE with accountability and for that they provide unique global identity to every user in system that helps to identify misbehaving user of PHR that gives decryption key to other unauthorized user.

Here the scheme trace that user who is misbehaving by his global identity, so burden of trust assumption on both side of authorities and PHR users. They provide analysis for scheme that shows the scheme is secure and efficient. Xhafa et.al [5] Scheme has supported policy that contains AND Gate for multiple attribute values with wild card. In Xhafa et.al [5] is good research work further on these scheme here we provide brief view of this scheme that contains five steps as given in fig 3.3.4.

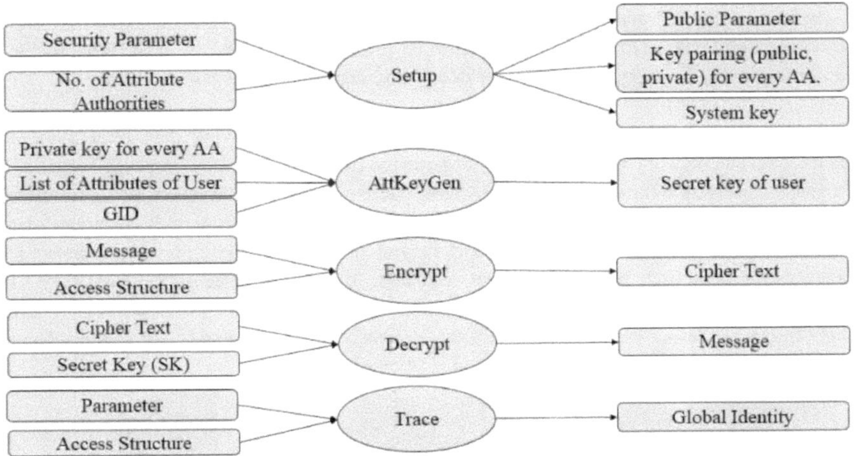

Fig. 3.3.4. Scheme analysis of Xhafa et.al [5]

Setup:
- Input: security parameter $\lambda \in$ N total number of Attribute Authority.
- Output: params as system parameters and N number of {public key, private key} pair.

AttKeyGen: run by every Attribute Authorities
- Input: private key of AA, list of attributes and global identity of user for which they created key.
- Output: decryption key according to given attribute list for user with unique identity.

Encrypt: run by PHR owner
- Input: massage and policy for encryption to generate ciphertext where policy contains some attribute that are subset of total attribute.
- Output: encrypted data ciphertext with respect to access structure.

Decryption: run by PHR user
- Input: ciphertext that is encrypted with some access policy and secret key of PHR user according some attributes.
- Output: they gets original message or not on bases of what attributes they have and it's satisfy the access policy that embedded in ciphertext or not.

Trace:
- Input: public parameters, cipher text policy
- Output: global identity of misbehaving user.

In Xhafa et.al [5] analysis part of scheme they claims that the scheme is secure under Decisional bilinear, Diffie-Hellman Decisional-linear Assumption. In their scheme they consider timing for encryption as and for decryption step timing is as where

In Dubovitskaya et.al.[14] author has mainly focus on achieving data integration on heterogeneous network within boundary of patient's privacy. First, they proposed secure and scalable architecture for storage and exchanging patient's health data. Second, an algorithm for effective aggregation of patient's health data for aiming to research from multiple independent sources. But for getting these functionalities they identifies problem on previous schemes for aggregating data from multiple sources. And then for successful achieving goal they use pseudonyms by multi-key searchable encryption and generalization by binary tree. However they shows that only ABE is not sufficient, so they use multikey searchable encryption scheme. Proposed architecture of Dubovitskaya et.al. [14] as shown here.

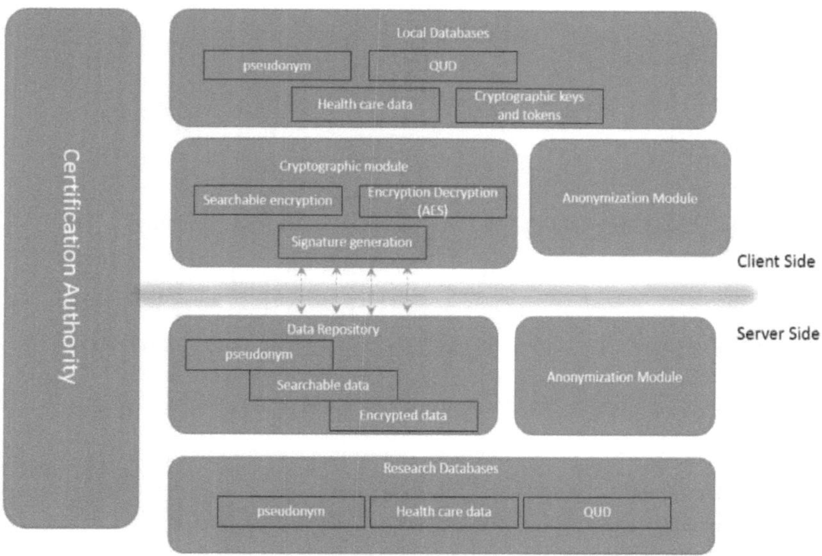

Fig. 3.3.5. Architecture overview of Dubovitskaya et.al. [14]

As shown in figure 3.3.5

there is mainly three parts databases both side client and server, cryptographic module on client, Anonymization module on both side and a single certificate authority.

Databases: Local databases (LDB) are store on patient side and contains health data. Data repository (DR) are hosted on cloud and generated in different hospitals. Research Databases (RSDB) are databases that store on purpose of research. Cryptographic Module: there are 3 parts (1) multikey searchable encryption (2) encrypt EHR before uploading (3) to generate the signature to ensure data is authentic

Author use two-factor authentication for accessing health data. Central Authority CA is responsible for issue certificates for public keys and smartcard for storing private key that is protected by PIN.

In Fabian et.al. [15] author proposed new area for securing huge amount of data (Big data) of medical/hospital records on cloud by dividing into multiple cloud. Their architecture features for selective access authorization and they use cryptographic secret sharing on disperse data on multicloud for reducing adversarial capabilities of cloud provider. On their implementation part of work they calculated timing for ABE and secret sharing in their scheme. For their given detail here we added chart for those data.

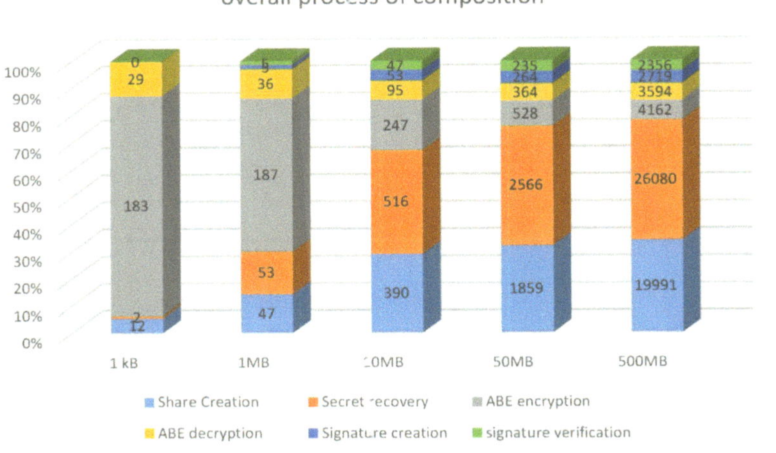

Fig. 3.3.6 Time composition of the overall process in Fabian et.al. [15]

3.4   Comparison

Table 3.4.1 Comparison of some schemes

Scheme	Parameter					
	Security	Access structure	Revocation	Key size	Cipher text size	Delegation
[1]Ibraimi. et.al.	Collision-Resistant	AND OR support	Yes	Secret-Key : no. of Attribute of user	Depend on Access policy	Yes
[2][3] Ming et.al	Collision-Resistant	AND, OR, Not ,Wild card	Efficient	No. of Attribute of user	(Set of Attribute for AA)+(Universal Attri.)	On-demand
[4]Alkinyele Et.al.	Collision-Resistant	AND,OR	Yes	Depend on Attribute	Depend on access structure	Yes
[5]Xhafa et.al	Collision-Resistant	AND, wild cards	Yes	Depend on Attribute	Depend on access structure	No

Table 3.4.2 Advantages and disadvantages of some schemes

Scheme	Advantage	Disadvantage
[1]Ibraimi. et.al.	Divide share-secret key given to user and mediator	Delegator must be Honest
[2][3] Ming et.al	Efficient and On demandUser Revocation,Break glass case,Multi Authority	MA-ABE not efficient.
[4]Alkinyele Et.al.	Created Iphone Apps,Interface with Google Health	Cipher text size overhead
[5]Xhafa et.al	Accountability, Give identity of misbehaving	Use of GID makes performance degradation, variable length cipher text size, variable secret key key size

4 Implementation of Bethencourt's cp-abe toolkit [17]

Here we show implement cp-abe toolkit on lab with configurations as Ubuntu 14.04 with 4GB RAM, Intel core i3 processer 3rd gen. Successful Working of CP-ABE toolkit for following scenario. There is two users: Sara, Kevin that needs to encrypt file sample.txt and secure it using CP-ABE scheme that can be stored at untrusted server or in cloud where user's privacy is not compromise. For working of CP-ABE contains different keys as master key, pub_key, priv_key_sara, priv_key_kevin. In encryption process of file/data with some access structure where we consider following scenario that access structure is shown in figure 4.1. After complete successful encryption process the file is encrypted in sample.txt.cpabe that is only decrypt if decryptor has some threshold level of attributes. For decrypting file here only Kevin is able to decrypt with his attribute.

Fig. 3.4.1 Access structure for encryption of sample file.

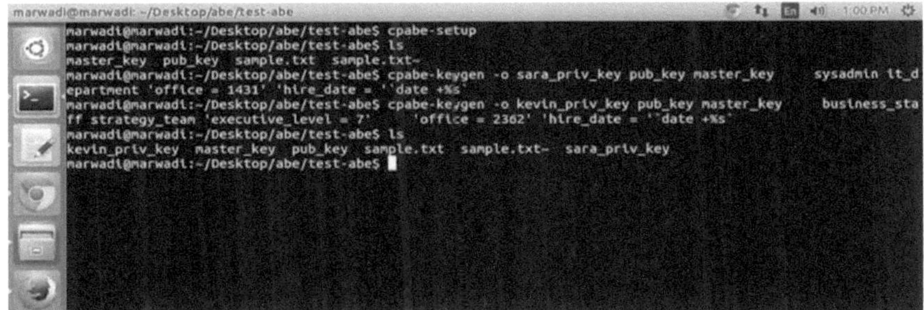

Fig. 3.4.2 Setup and key generation step on sample.text file

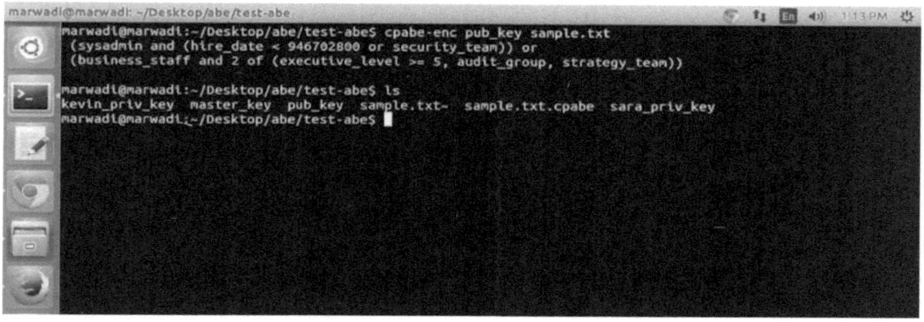

Fig. 3.4.3 Encryption of sample file using access structure.

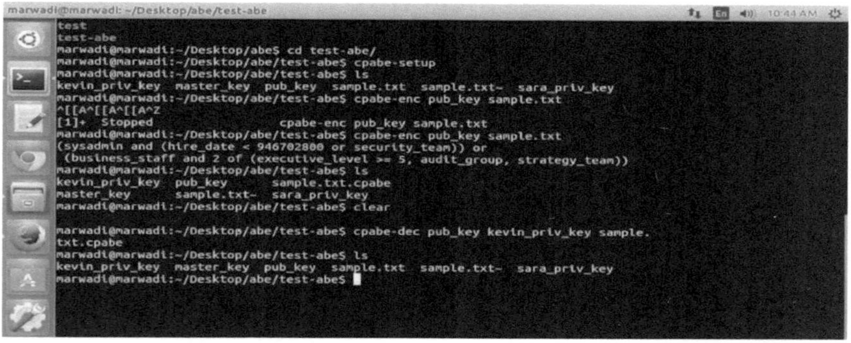

Fig. 3.4.4 Decryption of file using valid user

Fig. 3.4.5 Decryption of file using invalid user

5   Conclusion

Here we study some of the CP-ABE scheme that use for PHR that secure on cloud or untrusted data centers. We define comparison of different schemes in brief in this survey work. The future enhancement of field is to provide a single supreme scheme that provides all the functionalities that surveyed in this work. For adaptation of the CP-ABE in current days of smartphones need to work on cp-abe toclkit to work efficiently on android devices.

Acknowledgment

We are thankful to *Prof. Yogesh Ramani,* for helping us to solve typos and grammatical error throughout the book.

References

1. Ibraimi, L., Petkovic, M., Nikova, S., Hartel, P., Jonker, W.: Ciphertext-Policy Attribute-Based Threshold Decryption with Flexible Delegation and Revocation of User Attributes. (2009).
   Link: http://doc.utwente.nl/65471/1/Technical_Report.pdf
2. Li, M., Yu, S., Ren, K., Lou, W.: Securing personal health records in cloud computing: Patient-centric and fine-grained data access control in multi-owner settings. Lect. Notes Inst. Comput. Sci. Soc. Telecommun. Eng. 50 LNICST, 89–106 (2010).
   Doi: 10.1007/978-3-642-16161-2_6
3. Li, M., Yu, S., Zheng, Y., Member, S.: Scalable and Secure Sharing of Personal Health Records in Cloud Computing Using Attribute-Based Encryption. IEEE Trans. PARALLEL Distrib. Syst. 24, 131–143 (2013).
   Doi: 10.1109/TPDS.2012.97
4. Akinyele, J.A., Lehmann, C.U., Green, M.D., Pagano, M.W., Peterson, Z.N.J., Rubin, A.D.: Self-Protecting Electronic Medical Records Using Attribute-Based Encryption. s1st ACM Work. Secur. Priv. smartphones Mob. devices. 75–86, (2011).
   Doi: 10.1145/2046614.2046628
5. Xhafa, F., Feng, J., Zhang, Y., Chen, X., Li, J.: Privacy-aware attribute-based PHR sharing with user accountability in cloud computing. J. Supercomput. 71, 1607–1619 (2014).
   Link: http://hdl.handle.net/10945/40165
6. Sahai, A., Waters, B.: Fuzzy Identity-Based Encryption. Adv. Cryptol. – EUROCRYPT. 3494, 457–473 (2005).
   Doi: 10.1007/11426639_27
7. Bethencourt, J., Sahai, A., Waters, B.: Ciphertext-Policy Attribute-Based Encryption. Secur. Privacy, SP '07. IEEE. 321 – 334 (2007).
   Doi: 10.1109/SP.2007.11
8. Chase, M., Chow, S.S.M.: Improving privacy and security in multi-authority attribute-based encryption. Proc. 16th ACM Conf. Comput. Commun. Secur. 121 (2009).
   Doi: 10.1145/1653662.1653678
9. Benaloh, J., Chase, M., Horvitz, E., Lauter, K.: Patient controlled encryption: ensuring privacy of electronic medical records. CCSW '09 Proc. 2009 ACM Work. Cloud Comput. Secur. 103–114 (2009).
   Doi: 10.1145/1655008.1655024
10. Atallah, M.J., Blanton, M., Fazio, N., Frikken, K.B.: Dynamic and Efficient Key Management for Access Hierarchies. ACM Trans. Inf. Syst. Secur. 12, 1–43 (2009).
    Doi: 10.1145/1455526.1455531
11. Damiani, E., di Vimercati, S.D.C., Foresti, S., Jajodia, S., Paraboschi, S., Samarati, P.: Key management for multi-user encrypted databases. Proc. First ACM Work. Storage Secur. Surviv. 74–83 (2005).
    Doi: 10.1145/1103780.1103792
12. Wang, W., Li, Z., Owens, R., Bhargava, B.: Secure and Efficient Access to Outsourced Data. Proc. 2009 ACM Work. Cloud Comput. Secur.. 55–65 (2009).

Doi: 10.1145/1655008.1655016

13. Boldyreva, A., Goyal, V., Kumar, V.: Identity-based encryption with efficient revocation. Proc. 15th ACM Conf. Comput. Commun. Secur. - CCS '08. 417-426 (2008).
Doi: 10.1145/1455770.1455823

14. Dubovitskaya, A., Urovi, V., Vasirani, M., Aberer, K., Schumacher, M.I.: A Cloud-Based eHealth Architecture for Privacy Preserving Data Integration Alevtina. ICT Syst. Secur. Priv. Prot. IFIP Adv. Inf. Commun. Technol. 455, 585–598 (2015).
Doi: 10.1007/978-3-319-18467-8_39

15. Fabian, B., Ermakova, T., Junghanns, P.: Collaborative and secure sharing of healthcare data in multi-clouds. J. Supercomput. 48, 132–150 (2015).
Doi: 10.1016/j.is.2014.05.004

16. Waters, B.: Ciphertext-Policy Attribute-Based Encryption : An Expressive, E fficient , and Provably Secure Realization. Springer, Lect. notes Comput. Sci. 6571, 55–70 (2006).
Doi: 10.1007/978-3-642-19379-8_4

17. cp-abe library, libbswabe library" cpabe-0.11 , libbswabe-0.9 "
http://acsc.cs.utexas.edu/cpabe/

18. Kaelber, D. C., Jha, A. K., Johnston, D., Middleton, B., and Bates, D. W., A research agenda for personal health records. J. Am. Med. Inform. Assoc. 15(6):729–736. (2008).
Doi: http://dx.doi.org/10.1197/jamia.M2547

19. US Public Law, "Health Insurance Portability and Accountability Act of 1996," 104th Congress, Public Law 104–191. (1996).
Link: Health Insurance Portability and Accountability Act of 1996-hippalaw.pdf

Acronyms and Glossary

Acronyms	Meaning
AA	Attribute Authority is Authority that manages small set of Attributes
ABE	Attribute Based Encryption is technique for Encrypting plaintext with some attribute.
Accountability	Accountability means responsibility of user on sharing secret part of key
ADL	Attribute Delegation List is a list of delegates that use delegation of other users. This is managed by Mediator
ARL	Attribute Revocation List is list for list of revoked users that managed by Mediator
CA	Central Authority that manages Public Key And Private Key in PKE.
CP-ABE	Ciphertext Policy Attribute Based Encryption
CP-ABTD	Ciphertext Policy Attribute Based Threshold Decryption
DAAS	Database as a Service is service provided by cloud as database
DR	Data Repository generated by hospital and sited on cloud for Dubovitskaya et.al. [14]
EHR	Electronic Health Record
FIBE	Fuzzy Identity Based Encryption is extended part of IBE where identity are given as set of string
GID	Global Identity is unique identity of User
IBE	Identity Based Encryption is encryption technique based on PKE but Public key as identity of user
KP-ABE	Key Policy Attribute Based Encryption is one of the technique under ABE that Attribute is attach with key
LD	Local databases that are store on patient side and hosted in cloud for Dubovitskaya et.al. [14]
LSSS	Linear Secret Sharing Scheme
Mediator	Mediator is intermediate for that providing some of secret key and also verifies users.
Multi Authority	Multi Authority means multiple AAs monitors set of attributes
Multi cloud	Multi cloud means disperse data on multiple cloud for reducing adversarial capabilities of cloud provider
PCE	Patient Centric Encryption is scheme where patient handles his/her records
PHR	Personal Health Record is Health record managed by patient itself.
PKE	Public Key Encryption is encryption technique with two different key as Public Key and Private Key, one use for encryption and other for decryption
PKG	Private Key Generator that generates private key for user in IBE
RSDB	Research Databases are database on purpose of Research for Dubovitskaya et.al. [14]
SD	Security domains is subset of users for reducing burden of key management
SKE	Symmetric Key Encryption is encryption technique with same key for encryption and decryption
TA	Trusted Authority is third party to produce public, private key pair.

About the Authors

Mr. Mayur Oza received Bachelor of Engineering in Information Technology from Government Engineering College, Modasa under Gujarat Technological University (GTU), Ahmedabad, India in 2014. He is currently pursing Master of Computer Engineering in Computer Engineering from Marwadi Education Foundation Group of Institution (MEFGI), Rajkot, India under GTU. He is interested in Research on Cryptography, Network Security, ABE, CP-ABE, and CP-ABE in PHR. He is life member of Cryptology Research Society of India (CRSI), Kolkata, India.

Ms. Nikita Gorasia received Bachelor of Engineering in Computer Engineering from R.K. College of Engineering, Rajkot under Gujarat Technological University (GTU), Ahmedabad, India in 2012. She has completed M.E. from Marwadi education Foundation, Rajkot in 2015. Currently she is working in the Department of Computer Engineering at Marwadi Education Foundation, Rajkot since 2015. She is interested in Research on Cryptography.

Dr. Nishant Doshi is a faculty in the Department of Computer Engineering at Marwadi Education Foundation, Rajkot since 2014. His main research interests includes algorithms, cryptography and remote user authentication, information protection in general. He has completed M.Tech from DA-IICT, Gandhinagar in 2009 and Ph.D. from NIT Surat in 2014. Along with active researcher, he is Editor-in-Chief of journals like IJCES, IJECEE, IJME, IJMES, and IJSCE. He is rewarded as *Young Scientist* from Venus International Foundation in year 2015.

YOUR KNOWLEDGE HAS VALUE

- We will publish your bachelor's and master's thesis, essays and papers

- Your own eBook and book - sold worldwide in all relevant shops

- Earn money with each sale

Upload your text at www.GRIN.com and publish for free